All copy rights are with the Author.
Data retrieval by any means electronics,
paper, digital etc is not permitted

MOLES 01

After the firm belief that matter exists in the form of atoms, the next stage was to count the atoms. Though, individual atoms can not be seen with the naked eye. We can see few elements atoms only with electron microscope.

6.022×10^{23}

Gas Cylinder _____?????

How many molecules

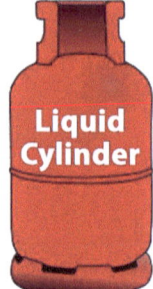
Liquid Cylinder _____?????

MOLES 02

AMEDEO AVOGADRO, a genius scientist of Itly was thinking to count atoms. This idea was bit weired for other scientists. But, Amedeo did not budge. Avogadro had experties in mathmatics and physics, this knowledge helped him to count the atoms.

He came to a solution that atomic weights and molecular weight can be imagined in grams.

Isn't it bit confusing!!

AMAZING FACTS
Avogadro started his career as a lawyer.
But he was more interested in sciences.
He studied sciences privately and become a college lecturer.

Yes, it sounds like that

Let's take the molecular weight of water

H_2O = 18 a.m.u

Here is the magic of Avogadro. Rather taking a.m.u. (Atomic Mass Unit), he considered it in grams.

The Property "grams" sounds a practical value to every one.

Alright !!!!
H_2O = 18 grams

Is there a good reason to express molecular weight in grams?

Yes, there is:
In the Lab, we measure everthing in grams either solid or liquid.

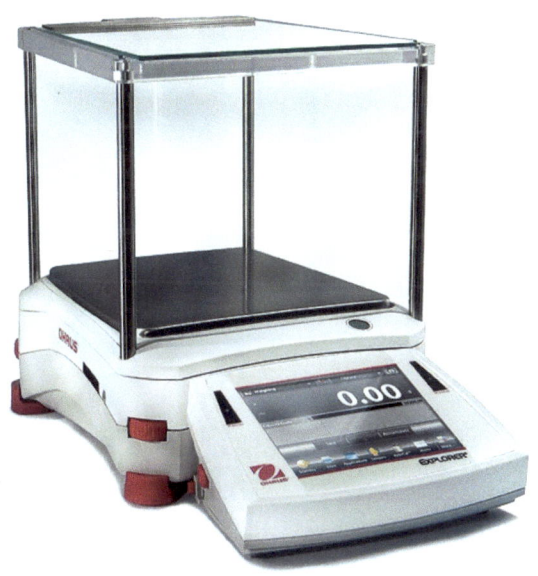

So, Avogadro reached to a best definition. "Atomic weight, molecular weight and formula weight expressed in gram mole."

MOLES | 07

"But, why "MOLE" was he found of that small, spiky animal??"

Mole is a collective term, as their is dozen for 12, century for 100 and silver jubliee for......................

MOLE = $\dfrac{\text{Weight in grams}}{\text{atomic mass, molecular mass, (Formula Mass)}}$

You can assume mass instead of the old term weight

Ok, I am not that much confused. I think Avogadro wanted to introduce new term Mole

MOLES | 09

So, what is Mole??

According to Avogadro, one Mole contains, 6.022 X 10^{23} particles. This number stretched to 23 noughts. These particles can be atoms, molecules and even electrons.

This number is called Avogadro's numbers

MOLES | 10

A glass of water

Spoonful of salt

A cylinder of carbon dioxide gas

We can calculate number of atoms in these materials.

Let's do some partical work.

John : How many moles are there in 2 grams of carbon dioxide.

Julia : Wait !!!
I have to use Avogadro's formula.

$$MOLE = \frac{\text{Weight in grams}}{\text{atomic mass, molecular mass, (Formula Mass)}}$$

Let us put values :

Mole : = $\dfrac{2 \text{ grams}}{44 \text{ molecular mass}}$

0.045 moles

Is it clear to you?
   No !!

Alright No.worrries !!

Let me have another approach.

How a coin machine works?
By electricty ! !

off course ! !
I mean, how does it count the coins
and give us currency notes.

Who tells the machine that there are different type of coins with varying shapes, size and weight.
wait a minute ! ! !

Yes, Size and weight matters

Actually, there are different trays in the coin machine.
Coins are assorted and move to the specific trays which are connected with digital weight machine.
Weights of all the trays are calculated and total amount is given in currency notes.

**Question :
What is some thing special over here?**

"The weight of the differnet coins"

We will apply this analogy with weight of different atoms.

Oxygen = 1/2 0 = 16

Carbon = C = 12

Hydrogen = H = 1

WHERE DID AVOGADRO LIVE?

Avogadro was born in Turine, Italy. His father was Noble. Later on he also become the Count of Quaregna. A count was mostly the companion of Emperor

Fact :
The great theory that gases are made up of molecules and molecules are made up of atoms, was given by Amedo Avogadro.

Some Mole Calculation:

How many moles are there in 3.2 grams of copper?

Atomic mass of copper
Ar (Cu) = 64.

By using Avogadro's Law Mass of one mole of cu = 64 grams.
3.6 grams of copper which will have

Moles = 3.2/64

= 0.05 moles of atoms

Fact :
The Complete name of Avogadro is Count Lorenzo Romano Amedo Carlo Avogadro

DO THESE MOLES HAVE ANY IMPORTNCE?

Yes,
Concenatrate on this equation

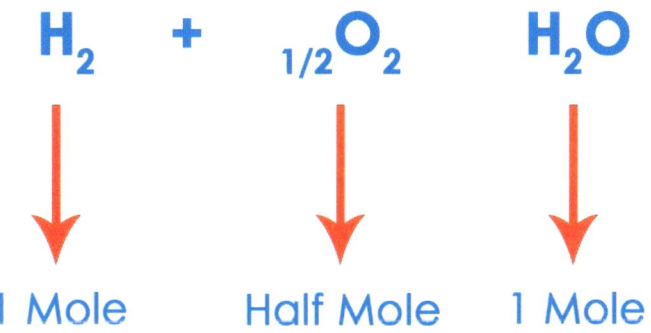

H_2 + $_{1/2}O_2$ H_2O

1 Mole   Half Mole   1 Mole

Ever before balancing, the mole values are clear.

Look carefully at arrow heads.

One mole of Hydrogen requires half mole of oxygen to one mole of warter.

When we know the quantites in moles, can central the quantities of reactants.

In the olden days, this was a common partice to put reactants by rough guess.

This worker is putting a roughly estimated quantity of reactant.

Thank God, the reactant is not exothermic.
If reactants are more exothermic, more quantities resut in exploision.

But, this scientist, Fritz Haber had blown up more than 100 cylinders to make ammonia.

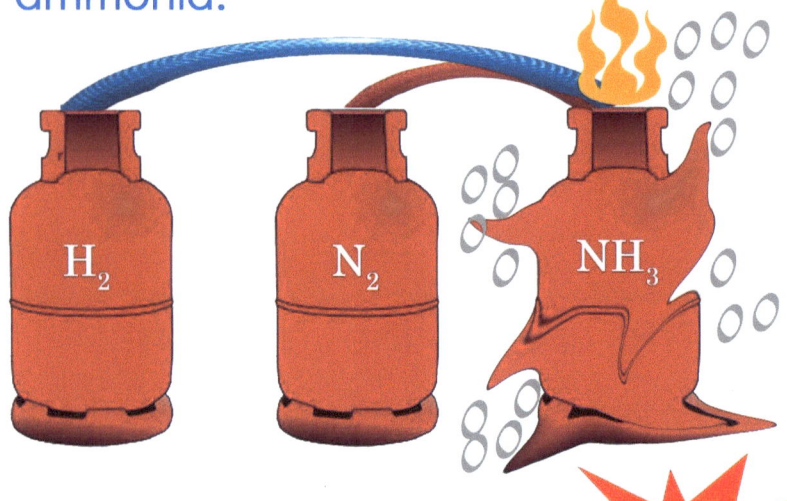

Though with corrected amounts of reactants, he succeeded.

HATS OFF TO AVOGADRO

Life saving medicines are made by counting atoms. Even if a single mole of atoms is in excess, a patient may die

Thanks Avogadro

Now get ready for some calculations

Example

How many moles of atoms should be there in 2.2 grams of copper?

Solution

The atomic mass for copper Cu is = 64.
1 mole of copper will weight 64 grams
Now,
   Using mole formula
   Moles of copper = 2.2g/64
                  = 0.034 moles

Yes, you have got the Chemical moles!!

Easy Questions:

Q.1 Find out the mass of
(a) 1 mole of chlorine atoms?
(b) 1 mole of chlorine atoms?
(c) 1 mole of iron atoms??

Q.2 How many moles of atoms are there in
(a) 10 g mole of calcium?
(b) 88 g carbon dioxide?
(c) 10 g of oxygen molecules?

MOLES AND GASES

Gases exhibit a different behaviour as compared to solid and liquids.

How Come?
We can not weigh up gases as we do with solids & liquids. When a gas is in a cylinder, it covers the entire bulk of the container.

Avogadro solved the problem that volume of a gas is dependent on the number of gas molecules, it dosn't a matter what is the nature of the gas.

So, here, Avogagro's moles work for us. Under standard temperature and pressure (S.T.P), gas of one mole = 22.4 dm³ (22400 cm³) But, at R.T.P (Room temperature and pressure), the gas of one mole is 24dm³.

ANSWER KEY

Ans. 1	Ans. 2
a 35g	a 0.25 mol
b 70g	b 2 mol
c 56g	c 0.625 mol

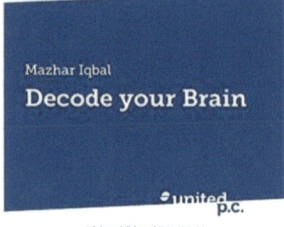

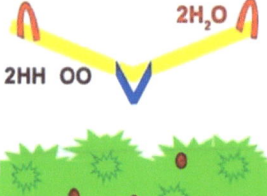

www.ingramcontent.com/pod-product-compliance
Lightning Source LLC
Chambersburg PA
CBHW041615180526
45159CB00002BC/865